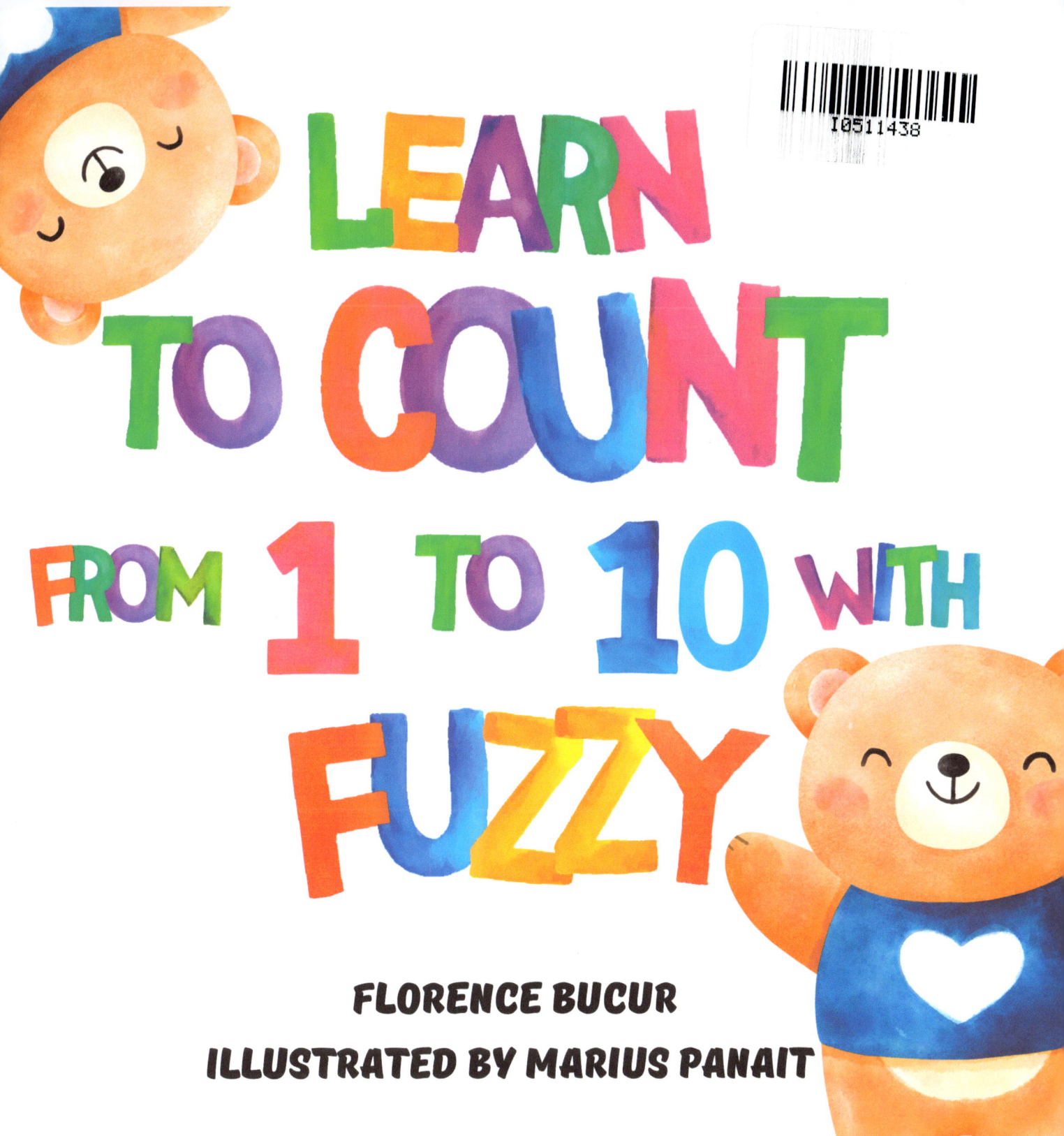

LEARN TO COUNT FROM 1 TO 10 WITH FUZZY

FLORENCE BUCUR

ILLUSTRATED BY MARIUS PANAIT

Let's count with Fuzzy
by Florence Bucur and Marius Banait

Text and illustrations © Copyright 2022
All rights reserved.

Tomorrow is Bobby's birthday, Fuzzy's little brother. He will be five years old.

Fuzzy is thinking of throwing Bobby a surprise party, but he needs his parents' help.

Bobby is visiting his grandmother. He has no idea what is going to happen.

Shopping List
- one cake
- two confetti bags
- three gift boxes
- four party banners
- five candles
- six party streamers
- seven party blowers
- eight party hats
- nine cupcakes
- ten balloons

Before going shopping, Fuzzy makes a list. He wants to make sure he doesn't forget anything.

Fuzzy and his parents arrive at the supermarket.

First of all Fuzzy has to buy a cake. He knows that Bobby really likes blueberries.

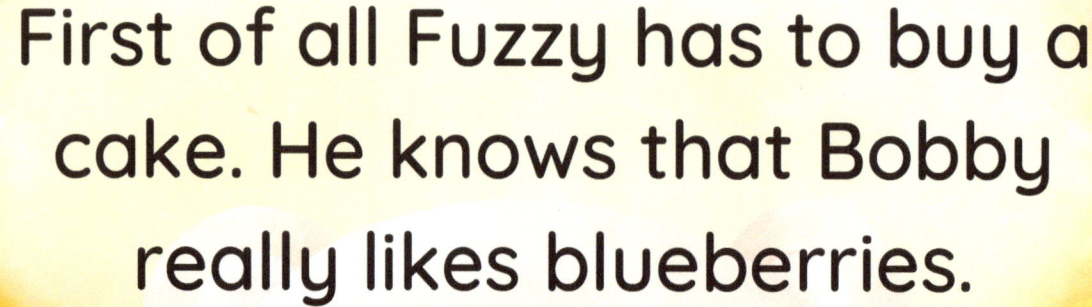

Now Fuzzy needs two bags of confetti.

Can you count the bags of confetti with him?

Congratulations!

Mom, Dad and Fuzzy need three boxes for presents.

-Do you think you can help them count the boxes?

-Congratulations, you did very well!

Next, Fuzzy has to buy four party banners.

Can you help him count four party banners?

You're doing very well!

Now the little bear has to buy five candles for the cake.

-Can you help Fuzzy count five candles?

1 2 3 4 5

-Excellent!

Next on the list are the six party streamers.

=Can you and Fuzzy count the six party streamers?

=Congratulations!

Fuzzy needs seven party horns.

Now he needs eight party hats.

Fuzzy has to buy nine cookies for the party.

-Can you help him count the nine cookies?

-You're doing very well!

He has to buy ten balloons.

Do you think you can help Fuzzy count ten balloons?

You did it! Now you can count to ten!

Time passes and after they finish shopping, Fuzzy has to send the invitations to Bobby's friends!

Tonight, Fuzzy has to decorate the backyard. Everything has to be perfect!

The big day has arrived! Fuzzy is excited for his brother to get home. He and his parents are waiting for Bobby.

Grandma blindfolds Bobby and leads him to the backyard.

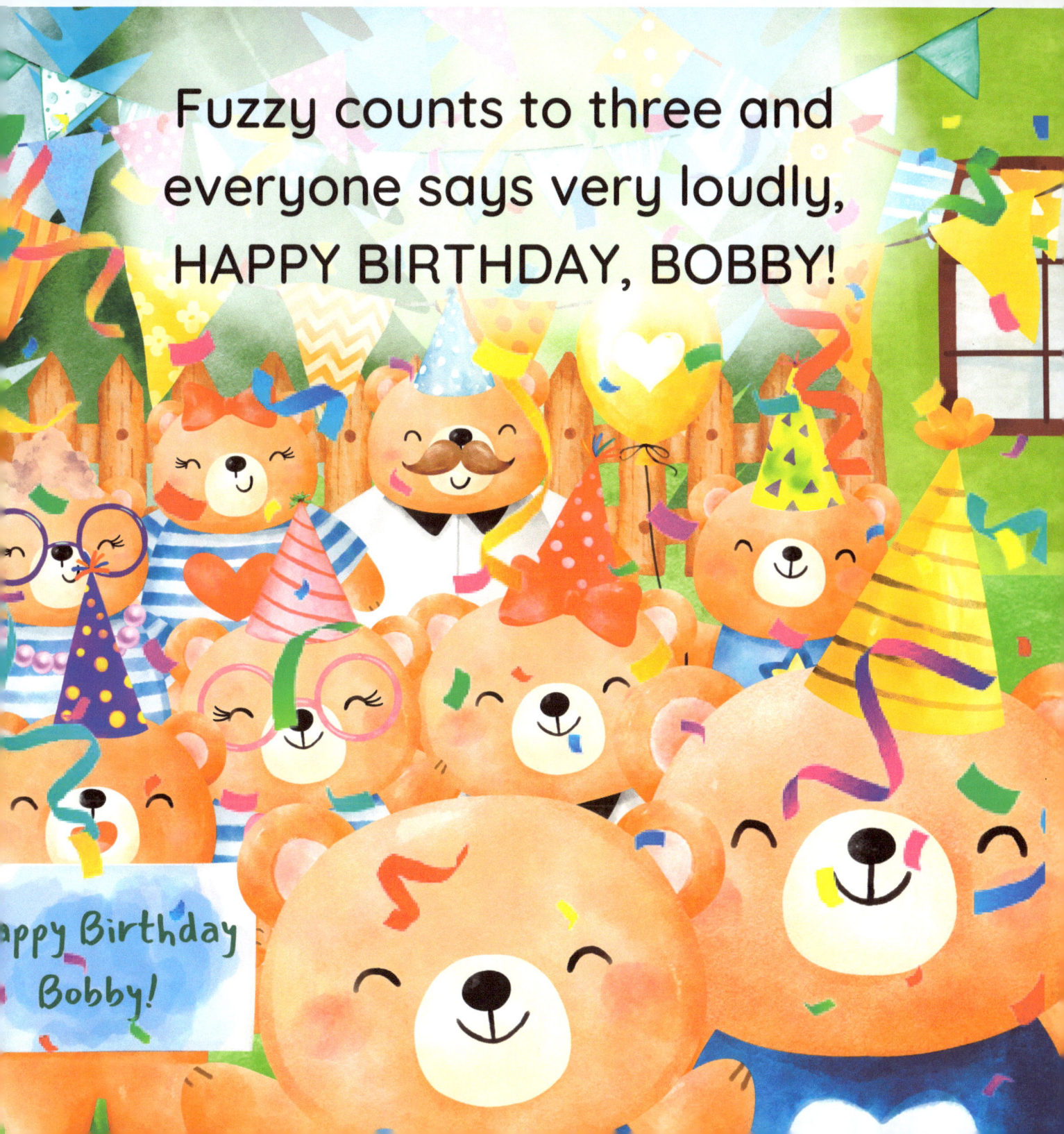

www.ingramcontent.com/pod-product-compliance
Lightning Source LLC
Chambersburg PA
CBHW051936210526
45473CB00006B/2275